W9-ATN-272

WHAT
IS A
VORTEX?

WHAT IS A VORTEX?

Sedona's Vortex Sites
A Practical Guide

DENNIS ANDRES

META ADVENTURES
SEDONA ARIZONA

Meta Adventures Publishing
Sedona Private Guides
583 Circle Drive
Sedona, AZ 86336
928-204-2201
www.SedonaPrivateGuides.com
www.MrSedona.com

For book or sales info contact:
Dreams In Action Distribution
P.O. Box 1894
Sedona, AZ 86339
928-204-1560
salesorbookinfo@dreamsinaction.us

©2000 Dennis Andres / Meta Adventures Publishing
Photographs ©2000 by Larry Lindahl
All Rights Reserved.
ISBN-13: 978-09721202-0-3
ISBN -10 0-9721202-0-3
Library of Congress Card Catalog Number 2002108581

1st printing, December 2000
14th printing, May 2012
No part of this publication may be reproduced, stored or
introduced into a retrieval system, or likewise copied
in any form without the prior written permission of the
publisher, excepting quotes for review or citation.

Title page: Cathedral Rock soars above Oak Creek

Contents

Oak Creek meanders through its canyon, Giant's Thumb on horizon

A Wish for Your Visit

WELCOME TO SEDONA! Over a decade ago, I drove into Sedona for the first time. Winding through Oak Creek Canyon in my car, I was stunned by the scenery. Towering limestone cliffs and ponderosa pine trees gave way to red rock layers, protecting a clear and gentle stream. Though I had been fortunate to visit more than forty countries in my life, I had never seen a place with natural beauty such as this. In a way that made no logical sense, I felt at home in a place I had never been to before.

Nonetheless, I was skeptical when I first heard the curious word "vortex," used as an explanation for this feeling. I had little clue then that I would someday move to Sedona and take up the challenge of answering the area's most often asked question for myself. Returning to America after several more years abroad, I moved to Sedona to begin a new career as an adventure guide. Not content to settle for either scientific uncertainty or spiritual clichés, I took up the question, "What *is* a vortex?"

So what makes me qualified to answer the questions you may have about Sedona's energy? It is neither an expertise in science nor in metaphysics, though I have studied both seriously. What qualifies me is that I am a good guide and a close observer.

I am able to bring people to places where they can have their own experience, and available to see what happens.

Now, having escorted nearly one thousand visitors throughout red rock country, I am ready to share what has been learned. The experiences of these people provide us with a pool of knowledge which helps to answer the most commonly asked questions about Sedona's vortexes.

In a small book the idea is not to exhaust your curiosity, but to provoke it. Likewise, I hope this will lead you to create your own Sedona adventure, as a way to explore these questions for yourself. The more you develop your relationship with this land, the more it will reward you with special moments and magical experiences.

Here's to your adventure!

DENNIS ANDRES
Sedona, Arizona

Section One

SEDONA'S
VORTEX ENERGY

QUESTIONS
&
ANSWERS

Kachina Woman rock spire near the trailhead in Boynton Canyon

Sedona's Vortex Energy:
Questions & Answers

❖

What is a vortex?

The scale by which we measure the aliveness of something or someone is called *health*. That's what we call people when they are full of life: *healthy*.

The most obvious way we recognize health is through physical indicators. When we see a healthy person, we may notice the tone of the skin, the brightness of the hair or the sparkle in the eyes.

A second sign of good health is that a person has what all healthy beings have: more energy. We can't necessarily point to a body part that contains the energy, but we know it is there.

This sounds fine, but what does it have to do with a vortex?

Consider one of the most significant advances in scientific thought of recent time. Having reviewed the cycles and changes in the geological, atmospheric and climactic aspects of our planet, scientists have begun to consider that the Earth itself is a living organism. Ironically, this is exactly what ancient peoples who have lived close to the land have said for centuries, and what environmentalists have been suggesting for decades. The Earth is alive.

This raises a couple of interesting questions. If the Earth is alive like other living beings, is it possible that there are places where it is healthier? If so, what would such places look like, and what would they be called?

Such a place would be called a *vortex*, and Sedona is one such place. This is the first part of the definition of a vortex:

A vortex is a place in nature where the Earth is exceptionally alive and healthy.

Welcome to Sedona, a very healthy place on planet Earth.

How do we know Sedona is a vortex?

Just as there are physical reflections of health among people, Sedona's beauty offers an indicator of its aliveness. For one thing, the light and color here grabs our attention. Sandstone formations of red, beige and orange rise above green forests of juniper, piñon and cypress beneath an azure sky. Just as a person in optimal health is more attractive, a vortex has a natural appearance that we find beautiful.

This is the second part of the definition of a vortex:

In a vortex the aliveness and health of the Earth is reflected in a tremendous natural beauty created by the elements of land, light, air and water.

Is Sedona a vortex just because it is beautiful?

Although we recognize physical aspects as indicators of health, we know that they are not the aliveness itself. So too the natural beauty here suggests something special is here, but it is not the vortex itself.

Recall that there is something else present in a state of robust health: more energy. A healthy person has it, and so does a vortex, suggesting the third part of our definition.

A vortex is a place on the planet of increased energy.

Ironically, when we're healthy it can be difficult to describe the energy that we possess other than as a feeling. We can't point to one part of our body where the energy is stored, and no X-ray reveals it.

In fact, it is difficult to measure energy in any way until after it has been utilized.

We have the same difficulty noting the energy here in Sedona. Generally speaking, there are few traditional measures of energy (such as increased magnetic or electric readings) that prove the vortex is here. Nonetheless, we have other ways of demonstrating it.

How does the energy of a vortex act?

We know that energy in healthy people helps them to do more, whether as physical exertion, mental concentration or emotional expression. So what would the energy of a vortex do?

The energy of a vortex acts as an *amplifier*. An amplifier takes a signal or frequency and makes it stronger. That's understandable for home stereos, but how does a vortex affect people?

The energy of a vortex acts as an amplifier. When we are in Sedona, the energy will amplify—or magnify—what we bring to it, whether on a physical, mental, emotional or spiritual level.

What frequencies are going on inside us? Quite a few, actually. Thinking thoughts and feeling emotions, for example, have measurable frequencies of energy that come from within. When we think and feel, the synapses in the brain produce chemical reactions that in turn lead to electromagnetic energies. They may be no more visible from the outside than a stereo signal, but they are equally real.

In Sedona, this amplifier effect manifests in a number of different ways. It may augment your thoughts and intuition, allowing you to gain an unexpected insight. It may heighten your feelings, as evidenced by the ways people feel especially happy to be here. It may have a physical effect, such as causing a nagging pain to dissipate. Finally, it can have a profound spiritual effect, leading a person to gain a greater understanding of who they are or where they are going in life.

If a vortex acts as an amplifier, which things will it amplify?

It is important to recognize that a vortex doesn't make judgments, any more than your stereo speakers influence which radio station you tune into. A vortex magnifies what is in it without bias.

Why doesn't everyone experience the same effect? Just as different stations send out different signals, people have different frequencies. The same person may also experience Sedona differently from one trip to the next. From this perspective, it becomes clear that the function of a vortex is not as tough to figure out as what's going on inside the people who enter it. A vortex will amplify, but it is up to you what "frequencies" you bring to it.

For most people the experience of the vortexes of Sedona is of a beautiful place where it feels good to be. For those who seek a deeper experience, one can do work through meditation, ritual, prayer, etc. to adjust one's own frequencies. Spending more time in Sedona can also help a person to become more aware of the amplifier's impact.

The remaining pages of this book will help you understand how to enjoyably connect with the natural energy in Sedona. For now, we can answer the question, "What is a vortex?" by saying:

A vortex is a place where the Earth is at its healthiest and most alive. The remarkable natural beauty of the area is the physical indicator of the aliveness. More importantly, the aliveness shows up in an increased energy that is present. The energy acts as an amplifier, magnifying what we bring to it on the physical, mental, emotional and spiritual levels.

Is this the only way to define a vortex?

There are plenty of different approaches to what a vortex is. Considering that it is complex and has a different impact on each person, this is not surprising.

When trying to compare the various definitions, perhaps the best question to ask yourself is, "What is a vortex to me?" The simplest way to answer it is by examining the kind of experience you have in Sedona.

Is there a scientific explanation of vortexes?

There are two approaches to answering this question. The first considers what we can measure using scientific instruments. The second considers the experiential evidence of people interacting with vortexes.

The most promising scientific hypothesis has been that the mineral composition of the red rocks creates a magnetism that has an impact on people. Unfortunately, there are no consistently significant magnetic or electric readings in Sedona, and the ones that are here don't correspond with the sites where people feel the energy. Further, other vortexes in the world have entirely different mineral compositions.

Since traditional measures are inconclusive, what does the experience of people at vortexes suggest? In observing the experiences of nearly a thousand people in Sedona, it is clear that there is a more sensitive, more accurate measuring device of the energy that is here. It is *us*: human beings. As complex organisms with an ability to sense more than what can be seen physically, we are often able to detect this energy.

Specifically, the evidence is that people seem to be responding to what scientists would call *subtle energies* (and to what alternative healers might call *chi*, and to what mystics might call *life force*). They don't seem to fit into the traditional categories of electric or magnetic, but we know they exist.

But why do people respond differently to the energy? In short, because people are different, and their senses are too. Just as someone with good eyesight sees differently than someone who needs glasses, so too someone who senses energy more acutely will feel subtle energies more deeply.

Are the vortexes scientific? If by scientific you mean a strictly physical phenomenon that can be measured by traditional devices, the honest answer is, "Probably not." If by scientific you mean substantiated through the direct experience of many visitors like yourself, then the answer is, "Absolutely."

Are vortexes masculine, feminine, electric or magnetic?

Like "electric" and "magnetic," the terms "masculine" and "feminine" are used by some to help people identify the nature of the energy at each site. It is a sincere effort to explain what is often a difficult-to-describe sensation.

While I respect the experience of others who have given these labels, I would suggest that there's no need to get hung up on the terminology. Too often, people believe they must feel the energy as masculine or feminine to be doing it right.

Similarly, some people call certain vortexes electric, others magnetic, and at least one electromagnetic. Technically, this is inaccurate. It is my belief that these terms arose to help people logically grasp the nature of the energy here, which feels differently in different areas. When faced with a new concept, we naturally refer to what is familiar and what is understood—electricity and magnetism—to try and explain it.

In charting the experience of nearly one thousand visitors, however, I've found that individuals meditating together in the same spot will often sense the energy differently. Some will feel uplifted; some will feel relaxed and mellow. Who's right? Everyone is.

For this reason, trust yourself as an expert in detecting energy. That way you can have a personal experience that measures up to the only standards that count: your own.

How many vortexes are there in Sedona?

There are a number of favored spots in Sedona to connect with the energy here. Traditionally, most people agree to four major points: Bell Rock, Cathedral Rock, Boynton Canyon, and Airport Mesa.

What does the human evidence suggest? It is clear that people are having energetic experiences in places throughout Sedona. In addition to the "Big Four," for example, people often report sensing energy near the Chapel of the Holy Cross and also in the Schnebly Hill area. Because these two are so often mentioned, you may hear people say that there are six or even more vortexes. Beyond this, locals and frequent visitors have their own special places in nature where they can feel the energy tangibly.

This suggests that Sedona as a whole is a vortex, with the energy spread throughout it like a bowl. For this reason, I recommend going to whichever sites you feel drawn. Which strike you as more interesting or more beautiful? Make an effort to go deeper into nature than you have before, feeling a sense of gratitude and appreciation wherever you end up.

Which is the best vortex to go to?

The best way to answer this question is to ask yourself another one: *Which one feels best for me?* That isn't an off-the-cuff remark to fend off your question. Instead, it's a delightful way to begin trusting your own intuition. Once you get there, trust your intuition further to walk to the area that seems best to you. You'll be tapping into a wonderful secret about developing intuition: the more faith you start putting in it, the more it grows.

The closest of the traditional vortex sites is Airport Mesa, with Bell Rock perhaps the most conveniently located on the highway that heads south to I-17 and Phoenix. Though each is marvelously beautiful, Cathedral Rock is perhaps the most photographed and Boynton Canyon hosts the most popular hiking trail.

How do I find a vortex?

Reaching the most well known spots is fairly easy, especially if you've got a car. *(See the map section for specific directions to each one.)* But where do you go once you're there?

Remember that the attitude you bring is more important than finding the right spot to stand upon. I've often seen people climbing all over Bell Rock desperately trying to find the vortex. It's not uncommon to see an earnest person shouting to their partner, "Over here Myrtle! I think I found it!" It's a pretty funny sight.

The truth is, *if you will genuinely find a place to sit quietly and connect, then you can't miss it!* My advice is to go to the general area by following the directions. The Forest Service signposts won't be of any help, because they mark hiking trails rather than vortex sites. Instead, use the opportunity to trust yourself and your own intuition. Walk to a place that feels right to you.

How long will it take to drive to a vortex?

The driving time to get to any of the most popular vortex sites from the center of Sedona ranges from five to twenty-five minutes. If you just want to jump out, snap a photo, and drive on to the next one, you could probably touch base at several within an hour or so.

The real adventure, however, is to spend a little time at one vortex and have a more in-depth experience. Once you arrive, you can generally walk into the vortex area easily. Since you'll be on the rocks and loose soil, walking shoes or hiking boots are best, but not mandatory. If your time is limited (say, you've only got an hour or two), pick one of the vortexes to drive to, rather than trying to visit all of them. They're not close, and although the scenery is beautiful, how many spiritual experiences are you likely to have behind the wheel? Correct: not many. Get out and walk around a bit. Find a nice place to sit and do a meditation of your own design, or try one of those suggested in this book.

If you've got a few hours, and the desire, there are wonderful hikes at the vortex areas, as well as elsewhere in Sedona. Any can become a spiritual experience if you are open to it.

If you're wondering how long it will take to feel the energy, it is different for each person. While some people feel energy even as they are driving into town, for others a more focused effort is required. To practice, the visualization suggested on page 51 takes about five to ten minutes to do.

What should I bring to a vortex?

If you're going to visit a vortex in normal conditions just for a short time, there's nothing specific that you must bring along, except perhaps your sense of adventure. For those who want a deeper experience, then the main thing to consider is the weather, so that you can be comfortable.

Sedona's climate is mild and pleasant for most of the year, but the dry air makes drinking water a must. In the summer months, you may want to add sun block and a hat to your list. In the winter, wear layers of clothes to regulate your comfort throughout the day. Finally, it's easy to get around at the four major vortex sites, but elsewhere it's best to bring a hiking book with maps.

Am I going to have to climb up anything?

Only if you want to. Airport Mesa, Cathedral Rock, Bell Rock and Boynton Canyon each involve relatively gentle walks. For those with limited mobility, it is possible to do the visualization suggested in the book from your car, as you park at the start of any of the trails. (Another option is to head to the Chapel area, which has wheelchair access.)

If you want to hike and climb, then enjoy. There are opportunities for further exploration on the trails at any of the vortexes, with Cathedral Rock and Boynton Canyon offering the longest hikes.

Though you may see some people trying to climb to the top of Bell Rock, it is unnecessary to do so in order to connect with the energy. In fact, it's a bad idea, since it is nearly impossible to descend safely from the highest peak.

When is the best time of day to go?

It's so beautiful in Sedona that just about any time works well. During morning hours, you can enjoy the unique natural light casting long shadows. This is also a good time to see wildlife. You will notice in Sedona how the light beautifully changes throughout the day. By late afternoon, the rock formations begin to glow and then turn bright orange as the sun goes down.

Sunset is beautiful to watch from any of the major vortexes, except perhaps Boynton Canyon, where the canyon walls block the sun. A particularly beautiful and easy-to-get-to spot to watch the sunset is Airport Mesa. Follow Airport Road past the turn-off for the vortex site all the way to the top of the mesa. Park to the left and walk across the street for the panoramic view.

Caution should be taken if you plan to go into nature at night, because the only lighting is from the stars and the moon. It may be completely dark at the vortexes, so if you head onto the trail without a flashlight, you're liable to get lost quickly. If you're lucky enough to be here in the days just before or during a full moon though, you might find that there's enough light to find your way.

What do I need to know when I go?

There are two things to bring in your vortex backpack that are more important than any map or compass. I call them, "Respect for Your Self," and "Respect for the Elements." With them, you'll have a great adventure.

To properly respect yourself, I mean that you need to come with an open-mindedness and a willingness to trust yourself. You don't have to believe in a vortex: it is real. But unless you trust your own ability to sense, it may be difficult to tell what, if anything, is really happening. How else will you know if that ache in your stomach is an energetic opening...or a sign that it's time for lunch?

This is a *critical* point for your vortex adventure. We have been conditioned to believe that what we imagine is false. We worry that at a vortex we won't be able to distinguish what is real and what is made up. The assumption is that if we only imagine something, it is not real.

I'd like you to consider a different way of perceiving the imagination. Your imagination is one of the chief ways that the aliveness of the vortex will reach your consciousness. For that reason, despite all the conditioning, you must learn to trust what you imagine. Until we learn to see beyond colors and hear beyond sounds, our imagination is the best way to tap into the subtle energies present in Sedona.

You can further respect yourself by recognizing that no matter what your religious beliefs, you *are* a spiritual being. You have a physical body, mental thoughts and emotional feelings, but beyond this there is a spirit within you that survives beyond this lifetime. There is a part of you that is aware and intelligent, seeking the greater meaning behind it all.

As for respecting the elements, we can take a lesson from the wisdom of the native peoples of the world, by recognizing our inherent connection with the Earth. They knew that just as the wind is like our breath, the sunlight like our body warmth, the water like our blood and the land like our flesh, we ourselves are of the Earth. But more than this, they knew that the elements also have a consciousness. In acknowledging the beauty and power of these special sites, we give them proper respect.

Knowing this, you can encounter a vortex from a place of mutual respect: the greatest avenue for learning, cooperating and working together to create success.

What should I expect to happen at a vortex?

It often seems that visitors in Sedona fall into two groups. The first group consists of people who are nervous about what might happen. They may be cynical (thinking that this is hocus-pocus), or afraid (fearful that feeling the energy means making changes in their life that they don't want to make).

In contrast are those who, you might say, try too hard. They have perhaps done a great deal of personal growth work and seek to feel the energy "the right way." To them, experiencing the energy becomes a spiritual litmus test. Secretly, they think that if they can feel the energy, this will prove that they are more enlightened.

Whether you fall into one of these groups or not, relax. A vortex won't force you to make changes, and it won't prove that your spirituality is better.

Connecting with vortex energy is a unique experience for each person. Yes, there are some things in common about people's experiences, but we can't really predict who will experience what.

Since it is a personal experience, there's no need to measure what happens to you against what happens to others. Put simply, there is no single correct way to sense a vortex.

Knowing this, you can take the pressure off yourself and enjoy what happens.

How can I
feel the energy?

The visualization in the "Working with Vortex Energy" section of this book gives a simple but enjoyable technique to help you sense the energy directly. In the meantime, there are some ways to note the energy's impact on you. First take note of your feelings at the site (or in Sedona overall). Do you feel more at peace, somehow "at home"? Perhaps you feel more delighted and energized than usual. Second, pay attention to physical sensation. Is any part of your body feeling affected, even if it is just more comfortable? Third, notice the thoughts, memories or ideas that come to your mind.

These could be signs that the energy is having a subtle impact on you. People who have done some kind of energy work, such as alternative healing, meditation or yoga often feel something different about Sedona. They may note a heightened state of emotion. Others sense a tingle in their hands, an opening in the heart area, or a slight pressure at the crown of the head. Finally, many simply feel a rush of new ideas and dreams, insights on action to take regarding an issue back home.

Nonetheless, even if you have none of these tangible symptoms, don't judge your experience a failure. For most of us, it takes a more focused effort to connect to the potent and yet subtle energy here. Recognize that just because it is subtle, doesn't mean it is weak. It means it is not obvious to the standard five senses.

How does it
feel for most people?

In the overwhelming majority of experiences, the feeling of connecting with a vortex is enjoyable and positive.

Over several years, I have led nearly a thousand people in a simple visualization which asks them to imagine the energy manifesting before their eyes. The visualization doesn't dictate to people how they should feel it, so that they will have their own personal experience. What are the results?

By this technique, one common way people will grasp the vortex is as energy itself. Some will sense it as a spiral, but this is just one way. Others sense the energy moving upwards from beneath the ground to the sky, or the opposite. Others visualize that energy moving in wave patterns, much like a movement of an ocean.

Another common sensing of the vortex is as light or color. At Bell Rock, for example, meditators sometimes perceive a white light around the rock, as if it had its own aura (which it does). Others instead will sense colors like red, or gold or green. Some will see more of a rainbow of color.

A third variation is for people to sense the energy as tone, vibration or sound. A few have mentioned hearing a word spoken, almost like a whisper, such as "Peace," or "Solitude."

A fourth possibility is to sense a new structure, formation, geometric or picture emerging in place of the red rock. While some have mentioned sensing a pyramid figure, a Mayan temple, or the impression of an ancient Indian *kiva*, others will sense a bubble of light, or a giant black hole.

Along with this mental experience, people will sense a gentle physical or emotional impact too. Some people feel their bodies as more relaxed, while others will feel a tingle of energy in their hands or some other part of the body. Most notice a subtle shift in their mood, as they become more peaceful and content.

This is the feedback from people spending just a few minutes to connect. There are much deeper, life-changing impacts for those who do more in-depth work.

Be open to the above phenomenon or any other sensation you may have.

Can it be an uncomfortable experience?

In an extremely small percentage of cases, being at a vortex is uncomfortable for people.

Many people have a strong emotional reaction. The vast majority of these are pleasantly emotional (they may cry tears of joy, for example), but there are exceptions where the experience is of sadness or grief.

A few people will have an odd physical response in their body. For example, one woman said she felt off-balance, as if she were tilted to the side. Another reported a headache; a third mentioned a discomfort in her solar plexus.

Why is this happening? Here's an example that may explain. Some time ago a visiting tourist from Germany came to my office and insisted that a vortex had made her very sad. Curious, I asked her to tell me about the current issues in her life. She told me that she had just left Germany a week before, and left behind her two young sons, who she would not be able to see again for four months.

"How does that make you feel?" I asked. Her response was immediate: she burst into tears. As we spoke, she came to see the amplifier effect of the vortex, bringing to the surface emotions that she had tried to shove down.

There is a happy ending to the story. Not only was she able to gently process the pain, but she was also able to go and work with these wonderful amplifiers to ask for healing. Additionally, in learning that energy which we cannot see with our eyes or touch with our hands is still real, she realized she could imagine sending energy to her boys. A simple technique I taught her also allowed her to sense their presence here. While it may not be a perfect substitute for their physical presence, it was comforting to her.

As for those with an uncomfortable physical reaction, none of the experiences I've seen have been severe. Interestingly, it seems that often the response is in a part of the body associated with the chakras (see page 61 for an explanation of what a chakra is). Thus the sense of the top of the head being pushed might relate to the seventh (crown) chakra, dealing with spiritual connection and the pain in the solar plexus with the third chakra, regulating emotion.

What are examples of the vortex effect?

Not long ago, I was out with a client who expressed her desire to see a tarantula. The critters don't inhabit her home state, and she wanted to see some of Arizona's wildlife. "Careful," I said to her, half-joking, half-serious, "because what you *think* gets amplified in the vortex." She responded quickly, "Safely...I wish to see a tarantula *safely*."

Tarantulas are around and about in the Sedona area, but they aren't commonly seen, especially on the major trails. So it was a great surprise to see something crawling along several minutes later. But this animal was much smaller than a tarantula, which is normally about the same size as a human hand. What could it be?

As it turned out, the tiny creature was a baby tarantula. "Safe," she said. "Ha, ha!" We were safe indeed, and let the spider pass. She experienced that thoughts—which are themselves measurable energies, as certain brain scanning devices can demonstrate—can be magnified in the energy of the vortex.

In another case, a client in his twenty's told me he was frustrated with his career. Although he enjoyed his work and was well-paid, he and his boss simply couldn't get along. While we walked in nature, he examined his own thoughts, beliefs and attitudes. He looked at past patterns that might be involved, as well as repressed emotions. Then I led him on a meditation to tap into the vortex energy and have it amplify whatever change he desired. I think he was skeptical, because there seemed to be no logical way out of the situation, without sacrificing his job.

Forty-eight hours later, after returning home to California, the client called. "Dennis," he said, "you'll never believe it! My boss just got transferred...and I got a promotion!"

One of the smallest miracles is also one that I have the most fun recalling. Recently, I worked with a family that included a young boy and girl. I tried to create a meditation active enough for the adults, but simple enough for the children. Before it began, I simply explained to the kids that they would have a chance to wish for something they wanted. What they didn't know was that the night before I had bought a small souvenir I intended to give them: dreamcatchers, a Native American handcraft placed over the bed at night.

I didn't ask anyone what they had wished for when they opened their eyes, but I reached into my backpack and handed the young boy the dreamcatcher. "Look Mom," he said, "Just what I asked for!"

Another couple, both suffering from chronic back pain, reported the disappearance of the pain during their stay in Sedona. A visualization I led them on at a vortex also led to a genuine physical sensation of energy in their bodies.

Individually, these experiences are anecdotal, and difficult to explain logically. But taken together, it would be illogical to deny the hundreds upon hundreds of magical occurrences. Also, as the accounts extend from health to business and career, relationship and creativity to financial abundance, I am convinced there is no limit to what can be amplified.

Why doesn't everyone feel the same effect?

The simplest answer to why people experience different effects is that people are different. Still, you might be wondering if there are any factors that account for some people having a deeper experience of the energy.

First, the length of time you spend here may have an impact. A short visit is pleasant, but has less impact than staying for a few days to spend time in nature.

Second, the level of intention and conscious awareness you bring matter. It seems that people willing to focus and concentrate on making a connection generally do better than those who wait for the effect to "hit them."

Third, the acuteness of your extended senses makes a difference. People who are in touch with their emotions, have developed their intuition and imagination, and have done prior metaphysical work (including energy healing such as Reiki and therapeutic touch) consistently have more potent experiences in Sedona. Just like the senses of sight, smell, taste, touch and hearing, the extended senses become more effective as you practice working with them.

Put it all together, and the effect is like jumping on a trampoline: the more spring you put into it, the bigger the bounce you receive. Drive quickly through Sedona and you may not feel much impact. But if you'll stay for a while, focus on connecting with the energy, and allow your senses to develop—wow, you'll feel it!

Who discovered
the Sedona vortexes?

You might say nobody in particular found or discovered them since they've been here all along. However, Sedonans credit Page Bryant, a spiritual channel as the first to use the term "vortex" in 1980, and Dick Sutphen, whose workshops and seminars helped many learn to connect with the energy. Each lives elsewhere in the United States and continues their spiritual work.

Are there other vortexes
in the world?

Although the term vortex has become associated chiefly with Sedona, it is not alone in being a place of power with an energy that amplifies. The planet has other such power places—sometimes called vortexes, sometimes not—whose energy also has this amplification effect. You've probably heard of some of them: Stonehenge in England, Machu Picchu in Peru, and Mt. Everest on the border of Nepal and Tibet. (Everest, by the way, is known to Nepalese as *Sagarmatha*, and Tibetans as *Chomolungma*, meaning, "Mother Goddess of the Earth." Sounds a lot more reverent than naming it after an 18th century British surveyor, don't you think?)

There is evidence of a vortex effect in the area of the pyramids of Egypt, and at the location of Mayan structures in Central America. In the United States, I would add Mt. Shasta in Northern California and areas of Hawaii to this list, among others.

The native peoples in these places sensed the energy that was present. In each society, there were those sensitive to energy—maybe the shamans, the medicine men, the high priests or priestesses. They tended to have a more direct connection with the Earth as they worked, lived and played upon it, aware of the subtle energies of the earth. They placed their massive stones or temples to mark the power sites.

Consciously or unconsciously, they knew that here their prayers, rituals and supplications to the divine would be *amplified*, and therefore, more likely to manifest.

Realizing that Sedona is not the only vortex in the world, you might reflect on a particularly special place you've been in nature somewhere around the world and ask yourself if it might be a vortex as well.

Did Native Americans know about vortexes?

Sedona has been home to a number of Native American peoples. The earliest group we know of were the Archaic people, who lived throughout the Southwest as far back as 5,000 B.C. The Sinagua (from the Spanish *sin agua*, meaning "without water") lived in the region approximately A.D. 600 until 1400. They built the marvelous ruins we find today at sites like Palatki and Honanki here in Sedona. It is believed that when they left the area they moved north to the Hopi mesas. Within the last two centuries, cultures such as the Yavapai have inhabited the Sedona area.

Did they know about the vortexes? It's a question which Native Americans answer differently. What we are certain of is that native peoples considered this sacred land, as evidenced by a Yavapai creation myth, which is associated with Boynton Canyon.

Ultimately the question is not one of vocabulary (as in, whether or not they used the word "vortex") but of the relationship to the elements and the Earth itself. Today we can recognize that a fundamental respect for the earth, wind, fire and water was an integral part of the life of those ancient ones who lived here.

What is the plural form of vortex?

There are two grammatically correct forms of the plural: vortices and vortexes.

Can I take a rock home from a vortex area?

In Hawaii, there is a book published full of the letters of folks who have taken rocks from the Kilouea volcano as souvenirs. Each of them details a sad story of bad luck that befell the unlucky tourists, mailed back with the original rock.

Perhaps a better idea than taking a rock is to create a memory that will last far longer as a souvenir. Take a moment while you are in Sedona to stand and look around. Then close your eyes. As you picture what you have seen in your mind, take a deep breath and smell the freshness in the air. Listen for the sounds of nature around you. Note the feeling in your heart and sense the relaxation in your body.

With this approach you can create a memory of Sedona that will last you until your next visit.

Section Two

FINDING THE VORTEX SITES

MAP
&
DIRECTIONS

Boynton Canyon is a favorite hiking area for many of Sedona's visitors

Finding the Vortex Sites:
Map & Directions

❖

Orientation to Sedona

Now that you are interested in investigating Sedona's energy, where should you go? The following section will help navigate your way through Red Rock Country. Directions are indicated from the intersection with a roundabout called the "Y," where Hwy 179 meets Hwy 89A. This point is often used as a reference for directions.

Highway 179 goes north-south to the Village of Oak Creek and Interstate 17. From where 179 meets the "Y," a right turn onto 89A leads northeast to Uptown Sedona and Oak Creek Canyon. This is also the route to Flagstaff and the Grand Canyon. A left turn onto 89A takes you to West Sedona. This is also a route to Cottonwood, the ghost town of Jerome and the historic city of Prescott.

Remember that since vortex energy is not strictly physical, there is no "X" that marks the spot where you'll find it. The energy is truly widespread throughout Sedona. The sites listed in this section are not the only places where you can connect with the energy, but they are the most well-known, and very beautiful.

Sedona has roundabouts rather than stop lights in several areas. Here are a few rules to follow when driving in a roundabout. If there is no car in the roundabout, just proceed. The vehicle in the roundabout has the right of way. YIELD, to the car in the intersection, as one can't always tell what direction that vehicle will be going in. To make a left hand turn, go around the roundabout first.

Red Rock passes are required to park in many trailhead areas. Some sites have vending machines that sell them for $5.00 per day, $15.00 per 7 consecutive days. For more info go to: www.redrock-country.org. For your safety: we ask that you stay on the trails, wear appropriate footwear and BRING WATER.

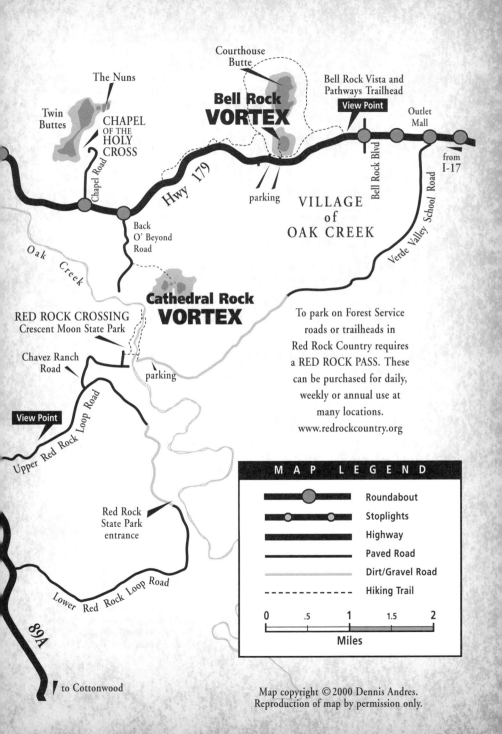

The Nuns

Twin Buttes

CHAPEL OF THE HOLY CROSS

Chapel Road

Courthouse Butte

Bell Rock VORTEX

Bell Rock Vista and Pathways Trailhead
View Point

Outlet Mall

from I-17

Hwy 179

Bell Rock Blvd

parking

VILLAGE of OAK CREEK

Back O' Beyond Road

Oak Creek

Verde Valley School Road

Cathedral Rock VORTEX

RED ROCK CROSSING
Crescent Moon State Park

Chavez Ranch Road

parking

To park on Forest Service roads or trailheads in Red Rock Country requires a RED ROCK PASS. These can be purchased for daily, weekly or annual use at many locations.
www.redrockcountry.org

View Point

Upper Red Rock Loop Road

Red Rock State Park entrance

Lower Red Rock Loop Road

89A

to Cottonwood

MAP LEGEND

⬤	Roundabout
● ●	Stoplights
	Highway
	Paved Road
	Dirt/Gravel Road
- - - -	Hiking Trail

0   .5   1   1.5   2
Miles

Map copyright © 2000 Dennis Andres.
Reproduction of map by permission only.

AIRPORT MESA

*The closest vortex to the
center of Sedona, this spot offers
views in all directions.*

What's best about Airport Mesa Vortex is what you can see from it. The closest vortex to the center of Sedona, the views are tremendous in nearly all directions. To the south and east are views of Bell Rock, Courthouse Butte, and Schnebly Hill. To the north you can see Capitol Butte (also known as "Thunder Mountain") with a ridge on its right ending at Coffee Pot Rock.

Be aware that it can be crowded here on weekends. The hike to the upper hill on the left is not for everyone. You may settle for splendid views at the saddle, about fifty feet uphill from the trailhead. Try early morning, or late afternoon, and afterwards you can continue up the road to the top of the mesa for a broad vista of Sedona itself and a peek at the airport.

DIRECTIONS: From the "Y," head west (away from Uptown) on Hwy 89A for 1.1 miles. Turn left onto Airport Road, and follow it for a half mile until reaching the small dirt parking area on the left.

Parking is limited (10-12) spaces and lot is small and uneven. No Red Rock Pass is needed to park here. If lot is full, go farther up the road, to city viewing area and more parking, while you wait for space in the lower area to open up. DO NOT walk down the very narrow 2-lane road. It is extremely dangerous and emphatically NOT recommended.

Twin Butte, Courthouse, and Bell Rock as viewed from Airport Mesa

The trail to Bell Rock is across solid sandstone

BELL ROCK

*Named for its shape,
this formation is easy to reach
and fun to explore.*

Bell Rock is a striking formation which everyone arriving to Red Rock Country from the south will see. Located along Hwy 179, access is easy. There are gentle walking trails to go for a stroll. You may be inspired to climb higher up, which rewards you with even better views. Use caution when going higher and standing close to ledges. The shale rock surface can be very fragile in areas. Avoid the very highest point though, which is steep and dangerous to reach.

The views from Bell Rock are always beautiful, but especially so in the morning. That's when the sun rises over the Munds and Lee mountains, warming you and lighting up the valley before your eyes.

DIRECTIONS: From the "Y" head south on Highway 179 for about 5 miles, or about fifteen minutes. There are two vista areas you can park in located on the left hand side of Hwy 179 as you approach the Village of Oak Creek; Courthouse Vista and Bell Rock Vista, the rock formations are clearly visible. Both have turning lanes so that you can cross over to the other side, watch for the signs after you cross Bell Rock Bridge.

Bell Rock Vista is a little further down the road and has a much larger parking area, the sign indicating the turning lane says Bell Rock Trail, turn here to cross, if you miss it, simply go around the roundabout and you will be guided to another sign indicating the turn. Red Rock Passes are required for both areas and are sold in machines on site.

BOYNTON CANYON

Offering the longest hiking trail
of any vortex site, Boynton Canyon lets you walk
at the foot of impressive crimson cliffs.

This is the longest hike available among the major vortex areas (up to three hours round-trip). The first thirty minutes of the walk takes you around Enchantment Resort, which may detract from the nature experience. Keep in mind that the tremendous popularity of this trail often means crowds in the spring and fall.

Some sense the energy strongest at "Kachina Woman," a tall red rock spire near the start of the trail. To reach it, follow signs for Vista Trail. There is more beauty deeper into Boynton Canyon if you choose to hike the full 90 minutes to reach the end of the trail.

DIRECTIONS: From the "Y", head west on Hwy 89A for 3.1 miles until the stop light at Dry Creek Road. It is approximately a 10-15 minute drive. Turn right and follow the road another 2.8 miles until you must turn left or right. Head left onto Boynton Pass Road toward Enchantment Resort, and follow for approximately 2.5 miles to the stop sign. Turn right, on Boynton Canyon Road and drive about 500 feet to parking area on your right. Red Rock passes are required and sold in a machine on site. Walk the Boynton Canyon Trail for about .03 mile, look for the sign in box and to your right for the sign to the Boynton Vista Trail.

A hiker enjoys the view at the end of Boynton Canyon Trail

Cathedral Rock reflects in the calm waters of Oak Creek in Crescent Moon State Park

CATHEDRAL ROCK

*This signature vortex is the most
photographed area in Sedona, with its red rock spires
towering over the flowing Oak Creek.*

In a town where scenic views are everywhere, Cathedral Rock stands out as a truly photogenic place. There are several approaches which take you up high among its spires or down below by the gentle Oak Creek. Bring your camera for wonderful sunsets which can turn the red rocks to a fiery orange color.

DIRECTIONS: There are three ways to get here, with the first and third bringing you to scenic Red Rock Crossing. Sedona's signature photo—of Cathedral Rock in the distance with the Oak Creek in front—is accessed via either the Upper Red Rock Loop Road or Verde Valley School Road routes.

Via Upper Red Rock Loop Road: The prettiest and smoothest drive is through West Sedona. From the "Y" head 4.1 miles through Sedona's west side to the stoplight at Upper Red Rock Loop Road. Turn left and enjoy the views as the road winds down nearly two miles. Turn left onto Chavez Ranch Road and follow it around the sharp curve for eight-tenths of a mile to the Crescent Moon State Park on the left. It has a paved pathway along the creek and lots of social (not official) trails. There is a fee for parking (under $10), not covered by the Red Rock Pass.

Via Back O'Beyond: A shorter route takes you to a more challenging trail which keeps you on the scenic west side, but out of view of the Oak Creek. From the "Y" head south on Highway 179 for 3.2

miles, turning right onto Back O'Beyond Road. Follow this road around its narrow curves until you see the trailhead parking lot on the left. Lot is small and no street parking is allowed. Red Rock Pass needed to park and is sold in machine on site. This uphill hike is hearty and gives you the best chance to get up among the spires of red rock.

Via Verde Valley School Road: A longer and somewhat rougher ride begins by heading south for 7.1 miles from the "Y" on Highway 179. You'll pass Bell Rock and 3 roundabouts. Turn right at the 3rd roundabout just past the Outlet Mall, onto Verde Valley School Road. Follow for four miles, to a generally well-maintained dirt road which most cars can handle, but is sometimes rough riding. Go slowly. Drive another mile and park in the lot on the left. A Red Rock Pass is required to park here and is sold at a machine on site.

Walk one hundred and fifty yards further along the road to reach Red Rock Crossing. You will pass a trail marker on the right, just keep walking down the road until you get to the turnaround and climb down over the rocks to get to Oak Creek and the crossing. Cathedral Rock is clearly visible to the east. You can walk a trail, down and to the right, to get closer to the formation itself.

Across the creek is the Crescent Moon Ranch Recreation Area. There is a $2 to $3 walk in fee on this side. To cross over, walk over to the stone footbridge that is to your left. Watch your step, as the rocks are slippery. Swimming across is an option if the temperatures are warm enough. This side of the creek lends itself to group outdoor activities. For more information call 928-203-2900, or go to www.redrockcountry.org.

CHAPEL OF THE HOLY CROSS

In a place of amazing
natural surroundings, only one
building matches it in beauty.

Perhaps no other building speaks better to the spiritual sense that the natural beauty here evokes than the Chapel of the Holy Cross. Built in 1956, it was the idea of Margaret Staude, who once wrote, "The doors of this chapel will ever be open to one and all, regardless of creed."

One wonders if Margaret was sensing a spiritual energy here, long before the term "vortex" was ever used. Standing at the plaza in front of the chapel and facing its doors, you can look to the left (east) to see two red rock pillars known as "The Nuns." To their left, a thin pillar of red rock stands. When Margaret saw it, she thought it looked like an image of Mary holding infant Jesus. At that moment, she decided to build her church on this hillside, in view of the rock now called "Madonna and Child."

The Chapel is open to visitors every day from 9am until 5pm, when the parking lot gate closes. For more information call 928-282-4069 or go to www.chapeloftheholycross.com.

DIRECTIONS: From the "Y," head south on 179 for 2.8 miles. Turn left onto the roundabout at Chapel Road and follow six-tenths of a mile to the Chapel and to parking area provided.

SCHNEBLY HILL

*If you can put up with a very
bumpy road, Sedona's historic hillside offers
interesting views and trails to explore.*

Adventurous drivers may take Sedona's most rugged road up to dramatic red rock plateaus offering stunning views. The road is not in much better condition than it was when finished in 1902, when it was the only road to Flagstaff. In that same year, T. Carl Schnebly, who had arrived here from Missouri, wrote to the U.S. Postal Service to get a post office for the area. He suggested "Schnebly Station" for a name, but it was rejected as being too long. His brother Ellsworth suggested instead that T. Carl name the town for his wife. He agreed and submitted the name you know today: Sedona.

This area also has spectacular views and rock formations that are referred to by the locals as Cow Pies. They provide a larger surface area to walk or meditate on.

DIRECTIONS: From the "Y" head south on Hwy 179 for .03 of a mile, you will cross a bridge and then drive in the roundabout. Take 2nd right onto Schnebly Hill Road and drive for about a mile until you reach the end of the paved road. Pull into the Huckaby/Margs Draw trail parking area on the left. A Red Rock Pass is required and is sold at a machine on site.

There are three trails you can take from here. Take the Munds Wagon Trail to the Cow Pies Trail for awesome views. Driving further is not advisable: Driving a vehicle with proper clearance (Jeep 4 X 4) is necessary. Road is EXTREMELY rough and gets worse farther along.

Section Three

Working with Vortex Energy

Techniques
&
Exercises

Aerial view at sunset of Cathedral Rock

WORKING WITH VORTEX ENERGY: TECHNIQUES & EXERCISES

❖

How can I utilize vortex energy?

Since a vortex is an amplifier, there is a wonderful opportunity to take something you dream of, something you desire, and let the energy here act upon it in a way that can bring it into your life.

The first way to utilize the energy is to let it flow over you, by feeling the peace or excitement that the spectacular nature here engenders.

A second way is to tap into the energy. The visualization which follows on page 51 gives you that opportunity.

A third way is to honor that energy through a ritual. The ritual suggested below is one that you may enjoy.

A fourth way is to work with the energy to co-create something in your life you desire to add or release.

What if I have never meditated before?

The good news is that nearly anyone can meditate. Like other things in life, meditation generally improves with practice. So don't feel dismayed if you doze off, or your mind drifts, or you feel distracted. That's normal.

Before you assure yourself that you can't meditate, consider that how you're defining it may be inaccurate. While Eastern tradition often uses meditative techniques that involve centering and the use of silence, this is not the only way. If this form leads you to fall asleep or your mind to wander, then try a guided meditation from a tape.

Do vortexes open up my psychic abilities?

A vortex is an amplifier. For that reason it is entirely possible that you'll find your intuition heightened while you are here. You may have unusually vivid dreams, or find new insight into the issues of your life.

A woman from Mexico once came to my office to tell me that in Sedona she had a very strong sense of the presence of her brother who had died a few years before. "It was as if he were sitting right next to me," she said, obviously moved. Later that same day, a woman from Philadelphia shared that at a similar spot, she felt the energy of a loved one who she had lost within the past year. In another case, a gentleman in his late seventy's reported waking up in the middle of the night and sensing the presence of a beautiful angelic being watching over him.

Be open to wonderfully unusual things happening to you in Sedona. It's par for the course.

Is there a simple visualization I can use?

Here's a brief visualization you can try. It does not require prior meditative experience, or that you achieve a state of extreme relaxation. All it requires is that you close your eyes! It is detailed enough to help you tap into the vortex energy, but short enough that you can remember it. You can memorize it and then do it; or if with a partner, one can read it to the other.

At the nature site you've chosen, find a place where you can sit comfortably. Remember that since there is no right or wrong way to feel the energy, you cannot fail this exercise. Take a moment to become comfortable, relax and then close your eyes. Do your best not to worry about where you've been, where you're headed, who may pass by, or the noise in the background. Take a deep breath and let all that slip away.

I will describe the process as if we were at Bell Rock, but you can use it for whichever Sedona vortex you prefer, or for that matter, any power spot you know of in nature.

Visualization Exercise

Begin by imagining that you are sitting at the base of Bell Rock looking up at it. As you observe it, something surprising and unexpected happens. Matter turns to energy, form returns to essence and somehow the energy of the vortex becomes visible or tangible to you. Perhaps you imagine Bell Rock morphing and changing shape, or losing its shape altogether.

As you sense this, actively imagine what is happening. Perhaps it turns into an energy pattern in a spiral formation, or like ocean waves. Maybe instead you sense it becoming one or many colors or lights. Or perhaps it appears to you as tone, vibration or sound, words and whispers. For you, maybe it is that you don't see or hear anything, but instead you just feel something there.

Whether it is one of these sensations or something else, the most important thing you can do is to continue to trust what you are sensing. Trust. Even if it feels that you are making it up, let that be alright.

As you observe the vortex of energy, you realize that Bell Rock is not the vortex itself, just the physical form it reveals itself as. Likewise, you are also energy that has taken a form. You become aware that with concentration, you can also return to the energy that you are. So you allow your body to become flexible and fluid, letting light and air penetrate between your cells. Then let those cells stretch and separate. Then imagine your own field of energy, just beyond your skin, and allow it to expand. Continue until you sense yourself as energy.

Slowly you—as energy—move toward the vortex of energy. As you approach, the edge of your energy field touches the edge of the vortex field. Now overlap your energies, and move more deeply inward until you sense yourself at its center, inside it or in the middle.

Once inside, you recall that a vortex is a place where the elements are in harmony, and where the Earth loves Herself. There is love here. See how you feel it. For some people it is an uplifting, re-charging, energizing sensation. For others it is a calm, peaceful feeling of oneness and connection. Or perhaps you sense some other feeling or energy: that's fine. Just trust whatever you feel, no matter how subtle it seems.

Now go deeper, actually imagining yourself soaking in, breathing in, drinking in the energy that is here. Let it penetrate you completely, and imagine it going down into your very cellular structure, down to your DNA. Imagine it changing you, having impact upon you.

If you like, take a moment to give something back to this vortex. Imagine sending out a bit of your hope or happiness. Perhaps instead you visualize sending gratitude for the beauty that is here.

When you like, imagine yourself pulling out of the vortex. As you separate, know that the energy is within you now. This vortex is a part of you, something you can reconnect with at any time in the future, merely by repeating this visualization.

Slowly you pull away, watching as Bell Rock returns to its form, to its shape. Then find yourself back on the rock where you began, observing Bell Rock, and returning to your form. When ready, gently open your eyes.

Congratulations! You've just tapped into the energy of the vortex. Resist the temptation to judge what you were sensing or feeling as not good enough, or not powerful enough. As you practice attuning to subtle energy your experience will deepen.

Is there a ritual you would recommend?

Since a vortex is a place where the elements exist in harmony, here's a good ritual for paying your respect to nature.

Ritual Exercise

Begin by turning to face to the East. With arms opened wide, sense the movement of the air, and call upon the Wind. Take a deep breath and feel the love of the air. Realize that it flows like the breath within you.

Turn one quarter to the right to face the South, and call upon the Light or the Fire. Sense the warmth of the sun on your body. Let it activate the heat, the life force within you.

Next turn to the West, calling upon the Water. Recall the water that once flowed here and comes again with the summer rains. Realize that it flows like the bloodstream within you.

Then to the North you turn, calling upon the energy of the Land. Imagine as if the love of the earth were moving from the ground up through your body to touch your heart. Become aware that the minerals of this land are very much like those of your own flesh and bone.

More than you ever have before, know that you are indeed one with the elements.

Some Native Americans also hailed the directions of up, to Father Sky, and down, to Mother Earth. They would call upon the energy in front (representing the future), the energy behind (representing tradition) and finally the direction inside or within. Out loud or to yourself, call upon the elements and the directions and ask open-heartedly for the healing or creation you would like in your life. If you can do this at a powerful time of day, such as sunset or sunrise, it can be even more potent.

Is there a healing or manifesting meditation?

Of course, the range of meditations that you could try during your vortex adventure is infinite. Here's a condensed version of one that I've often used, and which others have used to great effect. With it, they've created physical and emotional healing; financial and business success; greater career fulfillment; and dramatic relationship improvement. Give it a try.

Meditation

Once you've taken a few minutes to become relaxed, imagine that you are walking along a trail amidst the red rocks. On the way, you take the time to reflect. You reflect on the past, and how far you've come in your life. You reflect on the present, and get a perspective about where you stand in your life. You reflect on the future, considering the many positive possibilities that lay ahead.

Continuing along, you begin to focus on that future further. Think of something you'd like to let go of, first of all. It could be anything. Maybe a habit to end or a pattern to break comes to mind. Maybe there's something you want to heal, or be forgiven for. Maybe it's something material or financial (such as debt) that you want to let go of. Mull it over.

Whether many things or just one has come to mind, eventually you pick one to focus on. As you dwell on it, try to get as specific as you can about what you want. Think also about the essence it would give you, in other words, how would it make you feel to have it?

You're surprised to notice ahead on the trail a small object. Maybe it is something of nature, such as a stick, stone or feather. Maybe it is something you wouldn't expect to find in nature. As you approach, pick it up, make a firm choice that you are ready to release this thing from your life or to receive it into your life. Include in your choice a willingness for this to happen without struggle and pain, with harm to none.

As you continue along the trail, you now sense yourself coming toward one of the major vortexes. Walk up to the foot of it, and prepare to pay your respect to the elements with a sacred ritual. You do this by turning first to the East, and calling upon the Wind. As you do, imagine a breeze blowing across your cheek. Turning to the South, you call upon the Light, or the Fire. Sense a ray of sunlight touching you with its love. To the West you turn, now invoking the Water. Perhaps you hear the trickle of a nearby stream, and think about the love that flows in it. Then turning to face North, you call upon the Earth. As you do, you feel the love within the land rising up from it, flowing through your feet, legs and chest until it touches you heart.

Turning back to the East to close the circle, you notice that the vortex of energy is manifesting before you. Whether you sense it as color, light, vibration or otherwise, you get ready to jump in. Bringing the object with you, you jump in!

As you do, you begin to rise into the sky or fall through the earth quickly. Rising higher or falling deeper, you sense yourself moving away into space, or into a deep black void. Sense yourself coming to the edge of time and space until you POP! out of it.

Outside of time and space, you focus on this object. You fill yourself with imagination, thinking about what it will look like to have this success. Then sense the images and imagination overflowing you, flowing waves of imagination in all directions out from you.

You then feel the emotions that you would have if the success were here. Feel them flow, and overflow out from you. Finally, you expect the success, and you flow the expectancy— even if it's hard to believe you'll have it—in all directions around you.

At first there is silence here, beyond time and space. Then you sense something coming. Waves! Huge waves! And you realize that your imagination has been amplified by the vortex. Waves of imagination wash over you and fill you with new pictures of what the success might look like. Right behind these, enormous waves of emotions wash over you, filling you with the wondrous feelings you would have. Finally, a tidal wave of expectancy washes over you, and you expect the success as if it were guaranteed, as if it were awaiting you back home right now!

If you haven't already, let that object go. Watch it float away like a hot air balloon. Then come back into time and space, floating back toward earth, or soaring upwards until you return. Fifty feet away, then ten feet away, and run a few steps as you land safely.

Sense the vortex returned to the red rock form that represents it. Walk the trail home, breathing in your connection with the elements and feeling the gratitude for the success that is coming your way.

Appendix

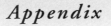

OTHER THINGS
YOU MAY BE WONDERING
ABOUT DURING YOUR
SEDONA VISIT

Twilight glows on Sedona's colorful sandstone formations under a full moon

❖

What does "New Age" mean?

Recognizing that "New Age" means many things to many people, here are a few of the core concepts of the philosophy that is so often mentioned in Sedona.

The term suggests there is something new going on. But what? People are changing, evolving in their consciousness and realizing that they can be responsible for creating a better life and a better world. Humans have it within their power to create a positive future.

One area where this power is demonstrated is through healing, and many New Agers recognize the value of alternative healing techniques. From acupuncture to organic nutrition to meditation, many approaches can fit in this category. They share the idea that healing has more to do with the intention and beliefs of the client than the medicine of the doctor.

A core idea of New Age thinking is that we need to establish a more balanced relationship with the planet. For many this means a more considerate approach to the environment, such as the treatment of natural settings, wildlife and the food we eat. In this way, the New Age often draws on the practices of ancient cultures which lived close to the land and in harmony with the Earth.

Another New Age concept recognizes the importance of the feminine. In considering God as a masculine energy only, society has cut off a source of power and nurturance that has filtered throughout our social relationships as chauvinism. Individuals have both a masculine and feminine side, and we can achieve our greatest success when both are acknowledged.

A fourth concept is that there is far more to our world than the physical reality we can see. There are angels and other spirit guides to assist us. Although a spiritual path is a matter of individual choice, it's not one we need to walk alone.

In this way all New Agers would be likely to agree that when you are clear in your intention and open in your heart, the universe has a way of lining things up to help you succeed. These synchronicities are a part of many Sedona adventures. Listen to your own inner guidance and you are likely to receive whatever assistance is needed to take the next step in your growth.

Are there crystals in Sedona?

Though there are plenty above ground in the stores, the truth is there aren't many under it. The crystalline formations seen in the red rocks are actually calcium carbonate.

Crystals and other precious stones are mined in other parts of Arizona, though. The clear quartz crystals that you see in the shops are typically from Arkansas, Madagascar or Brazil. Other stones come from around the world, from Russia to South Africa to Australia.

Many people in Sedona and outside of it believe crystals have their own "consciousness." Because of this, you can work with crystals as "partners" in healing and growth. Having their own properties and characteristics, there are crystals for enhancing relaxation, releasing emotions, and increasing intuition.

Like vortexes, you don't have to believe in crystals for them to work. But while you are here, you might buy one you're attracted to and notice how it affects you as you hold it through the day.

What are chakras?

Tradition passed down through the ages in Asia suggests that we have a series of energy centers that relate to different realms of our physical, emotional, mental and spiritual life. Though many say there as many as twelve chakras, all sources agree on at least seven.

The first chakra, associated with the color red, governs issues of safety and security. It is centered in the area of the tailbone. The second is slightly forward and down, behind the pubic bone. It relates to issues of pleasure and creativity and is identified with the color orange. The third chakra, in the abdomen, is yellow and relates to deep emotions. When people "get a feeling in their gut," this is where it is coming from. The fourth chakra governs issues of love and is called "the heart center." It is associated with the color green, though some people prefer the color pink. The fifth chakra, color blue, deals with all issues of expression and is centered in the throat. The sixth chakra is focused on wisdom and intuition, identified with the color indigo, and located in the center of the forehead. It is also called "the third eye," and you'll often see people slap themselves there when they remember something they had forgotten. The seventh chakra is at the crown of the head and is associated with the colors violet or white. It deals chiefly with our spiritual connection.

What's the story with past lives?

Simply put, a past life is a lifetime experienced in another time and place, other than in this body.

Consider that your soul is the part of you that exists both during and beyond lifetimes. It is the part of you that holds your deepest talents and gifts, and helps you to learn the lessons of love and growth that are best for you.

What does this have to do with Sedona? Well, it's not an absolute connection, but some people will have experiences while in Sedona that suggest memories of another lifetime. Coming to the Southwest may bring up images of a past life as a Native American, pioneer, cavalry soldier or missionary. Other times the amplifier effect will bring to the surface cellular memories about these lifetimes that while not making logical sense, feel like a *déjà vu* effect.

It may help you to find a facilitator who can lead you through a "regression," a meditative experience where you reconnect with the past life yourself. Although many will seek insight on fears or health issues in their past lives (such as discovering that you had once drowned as the explanation for a current fear of deep water), there are as many wonderful experiences that can be found.

Why investigate past lives? In exploring your past lives you may understand more of who you are, as well as your purpose on the planet this time around. Some would say that's the very definition of growth.

What is channeling?

Channeling is the process by which a person allows a spiritual guide to speak or work through them. Some channels work exclusively with one entity; others work with several. While a few channels are unaware of what the entity is saying as it is happening, most are semi-conscious. Conscious channels, a third type, will generally bring through the information and be aware of it at the same time, as if they were merely relaying to you what someone else in the room is saying.

There are many things in Sedona that come under the category of channeling. People who have attended a channeled session often find it a wonderful experience, as the spirit guide brings through a tremendous love and wisdom. But, not everyone agrees. If you try a channeled session while in Sedona, it is always worthwhile to check the information you receive against your own values and beliefs.

What is a psychic?

A psychic is someone who connects with energy and information that is beyond the standard senses and can share this insight to aid another person. Each psychic may have their own way of connecting, but generally they allow themselves a few moments to relax and "tune in" to your presence. In our energetic vibrations, we emanate information about ourselves, not only in the present, but also about our past and what we are headed towards in the future.

We all experience moments of intuition. Psychics, some by nature and some through training, have an enhanced sense of intuition which they can tap into at will.

Generally speaking, psychics prefer to serve as counselors offering guidance, rather than approaching a psychic session as a chance for them to guess what you had for lunch yesterday, it helps both you and the psychic to treat the session as an opportunity to gain insight on the key issues of your life.

Share Your Experience

What is your experience of the energy in Sedona?
Share how you feel and find out what others say on:
Mr. Sedona's Facebook and Twitter pages.

Visit Us

For more information on Red Rock Country, Vortex and other Apps, to reserve private guided outings, and find out more about upcoming events we will be offering, go to:
www.MrSedona.com.

About the Author

Dennis Andres is **Mr. Sedona**. He has lived and worked on four continents as a diplomat, management consultant and outdoor guide. Today, as an expert on energy and meditation, a certified yoga instructor, personal coach and local humorist, Dennis helps people make the most of Sedona. His books include the award-winning *Sedona's Top 10 Hikes*.

Fantastic Outings with Mr. Sedona

Don't settle for less than the best in Sedona. Escape the crowds for the silence and panoramas of the Red Rocks. For private and informative tours, hikes and vortex adventures, choose **Mr. Sedona** for singles, couples or families. Sign up for wonderful private group and corporate outings, or one-day and multi-day workshops combining health, energy, natural beauty and guided visualization. You'll love it: he guarantees it!

The USA Network, The San Francisco Chronicle, Every Day with Rachel Ray and many other media sources have featured Dennis Frommer's Budget Travel called him "a gifted guide and an endless resource." Call **928-204-2201** or check the website **www.MrSedona.com.**

Explore the Beauty and Magic of Sedona

Find out more about one of America's most beautiful places in these Best selling books by Dennis Andres:

Sedona's Top 10 Hikes
$11.95
An award winning publication. Sedona's most in-depth, beautiful and descriptive hike book.

What Is A Vortex?
$8.95
A practical guide to Sedona's vortex sites for inquiring minds.

To Order, Contact: Dreams In Action Distribution
e-mail: orders@DreamsInAction.us • web: www.DreamsInAction.us
phone: 928-204-1560 • write: P.O. Box 1894, Sedona, AZ 86339